Chapter 1: The Intersection of AI and Hydraulic Resource Management

The intersection of artificial intelligence (AI) and hydraulic resource management marks a significant milestone in the quest to preserve our planet's most vital resource: water. As populations grow and climate change intensifies, the pressure on water resources increases, necessitating innovative solutions to manage and conserve this precious resource. AI, with its capacity for data analysis, predictive modeling, and automation, emerges as a powerful tool in addressing these challenges.

AI technologies are being employed to monitor water quality and quantity in real-time. Sensors placed in rivers, lakes, and reservoirs collect data on various parameters such as pH levels, temperature, and contamination. This data is then fed into AI systems that analyze it to detect anomalies and predict potential issues before they become critical. For example, AI can predict harmful algal blooms, allowing authorities to take preventive measures to protect aquatic ecosystems and drinking water sources.

In agriculture, AI-driven smart irrigation systems are revolutionizing water use efficiency. Traditional irrigation methods often lead to water wastage due to over-irrigation or poor timing. AI systems, on the other hand, use weather forecasts, soil moisture levels, and crop requirements to optimize irrigation schedules. This not only conserves water but also enhances crop yields by ensuring that plants receive the right amount of water at the right time.

Moreover, AI is transforming the management of hydraulic infrastructure. Predictive maintenance, powered by AI, is becoming a game-changer for water utilities. By analyzing data from sensors embedded in pipes, pumps, and treatment plants, AI can identify signs of wear and tear, corrosion, or other issues that may lead to failures. This enables proactive maintenance, reducing the risk of costly and disruptive breakdowns while extending the lifespan of infrastructure.

Another critical application of AI in water management is in flood prediction and management. Floods are among the most devastating natural disasters, causing significant loss of life and property. AI models, trained on historical weather data and real-time information, can accurately predict the likelihood and severity of floods. This information is crucial for emergency response teams to plan evacuations, deploy resources, and mitigate the impact of floods on communities.

AI is also playing a role in water conservation efforts at the household level. Smart home devices, integrated with AI, help individuals monitor and reduce their water consumption. These devices provide real-time feedback on water use and offer recommendations for saving water. For instance, AI can suggest optimal times for watering lawns or highlight leaks that need to be fixed, thereby promoting water-efficient behaviors among consumers.

The integration of AI in hydraulic resource management is not without challenges. Data privacy and security are paramount, especially when dealing with critical infrastructure and personal information. Ensuring the accuracy and reliability of AI predictions is also crucial, as incorrect predictions can have serious consequences. Furthermore, there is a need for collaboration between technologists, policymakers, and water managers to develop and implement AI solutions effectively.

In conclusion, the intersection of AI and hydraulic resource management holds immense promise for the future of water conservation. By harnessing the power of AI, we can monitor and manage water resources more efficiently, predict and prevent issues, and promote sustainable water use practices. As technology continues to evolve, the potential for AI to contribute to the preservation of our planet's water resources will only grow, making it an essential tool in the fight against water scarcity and environmental degradation.

Chapter 2: Computer Vision for Water Quality Monitoring

Computer vision, a subset of artificial intelligence (AI), has made significant strides in various fields, including the preservation of hydraulic resources. Water quality monitoring, an essential aspect of water resource management, has greatly benefited from the advancements in computer vision technologies. By leveraging high-resolution cameras and sophisticated image processing algorithms, computer vision systems can detect and analyze contaminants in water bodies, providing real-time insights that are critical for maintaining safe and clean water supplies.

One of the primary applications of computer vision in water quality monitoring is the detection of harmful algal blooms (HABs). HABs are a major environmental concern as they can produce toxins harmful to humans, animals, and marine life. Traditional methods of detecting HABs involve manual sampling and laboratory analysis, which can be time-consuming and labor-intensive. In contrast, computer vision systems can continuously monitor water bodies using drones or stationary cameras, capturing images that are then processed to identify the presence of algal blooms. By analyzing the color, texture, and pattern of the water surface, these systems can accurately detect HABs and provide early warnings to authorities, enabling them to take preventive measures.

One of the primary applications of computer vision in water quality monitoring is the detection of harmful algal blooms (HABs). HABs are a major environmental concern as they can produce toxins harmful to humans, animals, and marine life. Traditional methods of detecting HABs involve manual sampling and laboratory analysis, which can be time-consuming and labor-intensive. In contrast, computer vision systems can continuously monitor water bodies using drones or stationary cameras, capturing images that are then processed to identify the presence of algal blooms. By analyzing the color, texture, and pattern of the water surface, these systems can accurately detect HABs and provide early warnings to authorities, enabling them to take preventive measures.

Another significant application of computer vision is in the detection of microplastics. Microplastics, tiny plastic particles less than 5 millimeters in size, have become a pervasive pollutant in aquatic environments. These particles can originate from a variety of sources, including cosmetic products, clothing fibers, and degraded plastic waste. Traditional methods of microplastic detection involve filtering water samples and examining them under a microscope, a process that is both tedious and limited in scope. Computer vision systems, however, can automate this process by capturing high-resolution images of water samples and using machine learning algorithms to identify and quantify microplastics. This technology not only accelerates the detection process but also provides more comprehensive data on the prevalence of microplastics in water bodies.

Moreover, computer vision can aid in the identification and tracking of invasive species in aquatic ecosystems. Invasive species can disrupt local ecosystems, outcompete native species, and cause significant ecological and economic damage. By deploying underwater cameras in rivers, lakes, and coastal areas, computer vision systems can capture images of aquatic organisms and analyze their characteristics to identify invasive species. This real-time monitoring allows for timely intervention and management strategies to control the spread of these species.

Despite the numerous benefits of computer vision in water quality monitoring, there are challenges that need to be addressed. The accuracy of computer vision systems depends on the quality of the captured images and the robustness of the image processing algorithms. Factors such as water transparency, lighting conditions, and camera resolution can affect the performance of these systems. Additionally, the deployment of computer vision technologies requires significant investment in infrastructure and maintenance.

In conclusion, computer vision has emerged as a powerful tool for water quality monitoring, offering real-time insights and automated detection capabilities that surpass traditional methods. By leveraging this technology, we can enhance our ability to monitor and protect water resources, ensuring the availability of clean and safe water for future generations. As advancements in computer vision continue, its applications in water resource management are expected to expand, contributing to the sustainable preservation of our planet's hydraulic resources.

Chapter 3: Deep Learning for Predictive Maintenance in Water Infrastructure

Deep learning, a subset of machine learning and artificial intelligence (AI), has revolutionized various industries with its ability to analyze vast amounts of data and identify complex patterns. In the realm of water resource management, deep learning is playing a crucial role in predictive maintenance of water infrastructure. By leveraging deep learning algorithms, water utilities can monitor the health of their infrastructure, predict potential failures, and implement maintenance strategies that minimize disruptions and extend the lifespan of critical assets.

Water infrastructure, including pipes, pumps, and treatment plants, is essential for the delivery of clean water and the management of wastewater. However, aging infrastructure, environmental stressors, and operational demands can lead to wear and tear, corrosion, and other issues that compromise the integrity of these systems. Traditional maintenance approaches often rely on reactive measures, addressing problems only after they have occurred. This can result in costly repairs, service interruptions, and even public health risks.

Deep learning offers a proactive approach to maintenance by analyzing data from various sources, such as sensors, historical maintenance records, and environmental conditions, to predict when and where failures are likely to occur. For example, sensors installed in water distribution networks can continuously monitor parameters such as pressure, flow rate, and temperature. The data collected by these sensors is then fed into deep learning models, which analyze the information to detect anomalies and predict potential leaks or bursts. By identifying these issues early, water utilities can schedule repairs before significant damage occurs, reducing water loss and service interruptions.

In addition to leak detection, deep learning is used to monitor the condition of water treatment plants. These facilities involve complex processes and equipment that require regular maintenance to ensure optimal performance. Deep learning models can analyze data from sensors that monitor chemical levels, turbidity, and other parameters in real-time. By identifying patterns and deviations from normal operating conditions, these models can predict equipment failures and recommend maintenance actions. This predictive maintenance approach not only improves the reliability of water treatment processes but also optimizes the use of resources and reduces operational costs.

Another critical application of deep learning in water infrastructure is the management of wastewater systems. Wastewater treatment plants process large volumes of water and handle various contaminants, making them susceptible to operational challenges. Deep learning algorithms can analyze data from sensors that monitor the quality and flow of wastewater, detecting anomalies that may indicate blockages, equipment malfunctions, or changes in influent characteristics. By predicting these issues, plant operators can take proactive measures to prevent disruptions and maintain compliance with regulatory standards.

Deep learning also plays a role in the management of stormwater systems, which are designed to handle runoff from rainfall and prevent flooding. These systems include a network of drains, pipes, and detention basins that can become overwhelmed during heavy rain events. Deep learning models can analyze weather forecasts, rainfall data, and sensor readings from stormwater infrastructure to predict flooding risks. This information enables municipalities to implement flood mitigation strategies, such as adjusting the operation of detention basins or deploying temporary barriers, to protect communities and infrastructure.

Despite the numerous benefits of deep learning for predictive maintenance in water infrastructure, there are challenges to consider. The accuracy of predictions depends on the quality and quantity of data available, as well as the robustness of the deep learning models. Additionally, the implementation of deep learning technologies requires investment in sensor networks, data management systems, and skilled personnel to develop and maintain the models.

In conclusion, deep learning has emerged as a transformative technology for predictive maintenance in water infrastructure. By harnessing the power of deep learning algorithms, water utilities can monitor the health of their systems, predict potential failures, and implement proactive maintenance strategies that enhance reliability, reduce costs, and ensure the continuous delivery of clean water. As the field of deep learning continues to advance, its applications in water resource management are expected to grow, contributing to the sustainable preservation of our planet's hydraulic resources.

Chapter 4: AI-Driven Smart Irrigation Systems for Water Efficiency

Agriculture is one of the largest consumers of water worldwide, accounting for approximately 70% of global freshwater withdrawals. With increasing water scarcity and the need to feed a growing population, improving water use efficiency in agriculture has become a critical priority. AI-driven smart irrigation systems have emerged as a promising solution to optimize water usage, enhance crop yields, and promote sustainable farming practices.

Traditional irrigation methods, such as flood irrigation and manual watering, often lead to significant water wastage due to over-irrigation and inefficient water distribution. Smart irrigation systems, powered by artificial intelligence (AI), use data-driven approaches to determine the optimal amount and timing of irrigation, ensuring that crops receive the right amount of water at the right time.

One of the key components of AI-driven smart irrigation systems is the use of soil moisture sensors. These sensors are placed in the soil at various depths to measure the moisture content in real-time. The data collected by the sensors is transmitted to an AI system that analyzes the information to determine the water needs of the crops. By continuously monitoring soil moisture levels, the AI system can make precise irrigation decisions, reducing water wastage and preventing over-irrigation.

In addition to soil moisture sensors, smart irrigation systems also incorporate weather data to optimize irrigation schedules. AI algorithms analyze weather forecasts, temperature, humidity, and rainfall data to predict the water requirements of crops. For example, if a significant amount of rainfall is expected, the AI system can delay irrigation, saving water and preventing soil erosion. Conversely, during dry and hot periods, the system can increase irrigation to ensure that crops receive adequate moisture.

Furthermore, AI-driven smart irrigation systems can be integrated with crop monitoring technologies to enhance water use efficiency. Drones equipped with multispectral and thermal cameras capture images of crops, providing valuable insights into their health and growth. The AI system analyzes these images to identify areas of stress, disease, or nutrient deficiency, which may require targeted irrigation. By applying water precisely where it is needed, smart irrigation systems can improve crop yields and reduce the overall water footprint of agriculture.

Another advantage of AI-driven smart irrigation systems is their ability to learn and adapt over time. Machine learning algorithms analyze historical data on weather patterns, soil conditions, and crop performance to continuously refine irrigation strategies. This adaptive capability ensures that the system becomes more efficient and effective with each growing season, leading to long-term water savings and improved agricultural productivity.

The benefits of smart irrigation systems extend beyond water conservation. By optimizing irrigation, these systems can enhance soil health, reduce the risk of waterlogging and salinity, and promote sustainable farming practices. Additionally, the precise application of water can reduce the need for fertilizers and pesticides, further minimizing the environmental impact of agriculture.

Despite the promising potential of AI-driven smart irrigation systems, there are challenges to their widespread adoption. The initial cost of installing sensors, drones, and AI infrastructure can be prohibitive for small-scale farmers. Additionally, the effectiveness of smart irrigation systems depends on the availability and accuracy of data, which may be limited in certain regions. Addressing these challenges requires investments in technology development, farmer education, and supportive policies to promote the adoption of smart irrigation practices.

In conclusion, AI-driven smart irrigation systems represent a significant advancement in agricultural water management. By leveraging AI and data-driven approaches, these systems optimize water usage, enhance crop yields, and promote sustainable farming practices. As water scarcity continues to pose a global challenge, the adoption of smart irrigation technologies will play a crucial role in ensuring food security and preserving hydraulic resources for future generations.

Chapter 5: AI in Water Resource Management and Allocation

Water resource management is a complex and multifaceted challenge that involves the allocation, distribution, and regulation of water resources to meet the demands of various sectors, including agriculture, industry, and domestic use. As populations grow and climate change impacts water availability, the need for efficient water resource management becomes increasingly critical. AI has emerged as a powerful tool in this domain, offering innovative solutions for optimizing water allocation and ensuring sustainable use.

One of the primary applications of AI in water resource management is the development of predictive models for water demand forecasting. Traditional methods of water demand forecasting often rely on historical data and simple statistical models, which may not accurately capture the dynamic and complex nature of water use patterns. AI algorithms, particularly machine learning models, can analyze large datasets, including historical water usage, population growth, weather patterns, and economic activities, to predict future water demand with greater accuracy. These predictive models enable water authorities to plan and allocate resources more effectively, ensuring that supply meets demand even during periods of scarcity.

AI also plays a crucial role in optimizing the operation of water distribution networks. Water distribution systems are vast and intricate, with numerous interconnected components, such as pumps, valves, and reservoirs. Managing these systems to minimize water loss, maintain pressure levels, and ensure reliable delivery is a significant challenge. AI-driven optimization algorithms can analyze real-time data from sensors installed throughout the distribution network to detect leaks, identify inefficiencies, and recommend operational adjustments. For example, by adjusting pump schedules and valve settings based on demand patterns and sensor data, AI systems can reduce energy consumption and water loss, resulting in cost savings and improved system reliability.

In addition to optimizing distribution networks, AI can enhance the management of water reservoirs and dams. Reservoirs and dams play a critical role in water storage, flood control, and hydroelectric power generation. The operation of these structures requires careful balancing of competing objectives, such as maximizing water storage, ensuring downstream flow for ecological health, and generating electricity. AI algorithms, particularly those based on reinforcement learning, can assist in developing optimal reservoir operation policies. These algorithms learn from historical data and simulations to identify strategies that achieve the best trade-offs between conflicting objectives. As a result, AI-driven reservoir management can improve water availability, enhance flood control, and increase the efficiency of hydroelectric power generation.

Furthermore, AI can support the development of integrated water resource management (IWRM) frameworks. IWRM is an approach that considers the interconnectedness of different water uses and aims to manage water resources holistically. AI technologies can facilitate the integration of data from various sources, such as hydrological models, climate projections, land use maps, and socioeconomic data, to provide a comprehensive understanding of water resource dynamics. By analyzing this integrated data, AI systems can support decision-making processes, identify potential conflicts, and propose strategies for equitable and sustainable water allocation among different users.

AI also has the potential to enhance community engagement and participation in water resource management. Digital platforms and mobile applications powered by AI can enable citizens to report water issues, such as leaks or contamination, in real-time. These platforms can aggregate and analyze citizen-reported data to provide valuable insights into local water challenges and facilitate timely responses from water authorities. Additionally, AI-driven tools can educate and empower communities by providing information on water conservation practices, promoting awareness of water issues, and encouraging behavioral changes to reduce water consumption.

Despite the significant potential of AI in water resource management, there are challenges to consider. The effectiveness of AI solutions depends on the availability and quality of data, which can vary across regions. Ensuring data privacy and security is also critical, as water infrastructure is a vital component of national security. Additionally, the implementation of AI technologies requires investment in infrastructure, capacity building, and collaboration among stakeholders, including government agencies, water utilities, researchers, and communities.

In conclusion, AI offers innovative solutions for water resource management and allocation, enabling more accurate demand forecasting, optimized distribution network operations, and improved reservoir management. By leveraging AI technologies, water authorities can enhance the efficiency, reliability, and sustainability of water resource management, addressing the growing challenges of water scarcity and ensuring the equitable distribution of this vital resource. As advancements in AI continue, its applications in water resource management are expected to expand, contributing to the preservation and sustainable use of our planet's hydraulic resources.

Chapter 6: AI for Flood Prediction and Mitigation

Floods are one of the most devastating natural disasters, causing significant loss of life, property damage, and economic disruption. Accurate flood prediction and effective mitigation strategies are essential for minimizing the impact of floods and protecting vulnerable communities. AI has emerged as a powerful tool in this domain, offering advanced capabilities for flood forecasting, risk assessment, and real-time response.

AI-driven flood prediction systems leverage a variety of data sources, including weather forecasts, satellite imagery, river flow data, and historical flood records, to develop accurate and timely flood forecasts. Machine learning algorithms analyze these datasets to identify patterns and correlations that traditional models may overlook. For example, deep learning models can process large volumes of satellite imagery to detect changes in water levels, soil moisture, and vegetation cover, which are critical indicators of flood risk. By combining this information with weather data, AI systems can predict the likelihood and severity of flooding with greater precision.

One of the significant advantages of AI in flood prediction is its ability to provide real-time insights. Traditional flood forecasting methods often involve complex hydrological models that require extensive computational resources and time. In contrast, AI algorithms can quickly process and analyze data, providing near-instantaneous flood forecasts. This real-time capability is crucial for early warning systems, enabling authorities to issue timely alerts and evacuate communities before floodwaters arrive. By reducing the lead time for flood warnings, AI-driven systems can save lives and minimize property damage.

In addition to prediction, AI plays a critical role in flood risk assessment. Understanding the factors that contribute to flood risk, such as topography, land use, and infrastructure vulnerabilities, is essential for developing effective mitigation strategies. AI algorithms can analyze geospatial data, including digital elevation models, land cover maps, and urban infrastructure layouts, to identify areas that are most susceptible to flooding. By mapping flood risk zones and assessing the potential impact on communities, AI systems can support decision-makers in prioritizing investments in flood protection measures, such as levees, floodwalls, and drainage systems.

AI also enhances the effectiveness of real-time flood response and management. During a flood event, timely and accurate information is critical for coordinating emergency response efforts. AI-driven systems can integrate data from various sources, including weather stations, river gauges, social media, and mobile applications, to provide a comprehensive and up-to-date picture of the flood situation. Machine learning algorithms can analyze this data to identify affected areas, estimate flood extent and depth, and predict the movement of floodwaters. This information enables emergency responders to allocate resources efficiently, deploy rescue teams to the most critical locations, and coordinate evacuation efforts.

Furthermore, AI can support post-flood recovery and reconstruction efforts. After a flood, assessing the extent of damage and prioritizing recovery activities is a complex and time-sensitive task. AI-driven damage assessment systems can analyze aerial and satellite imagery to identify damaged infrastructure, flooded areas, and debris accumulation. By automating the damage assessment process, AI systems can provide rapid and accurate estimates of the impact, enabling authorities to plan and implement recovery strategies more effectively. Additionally, AI can support the development of resilient infrastructure by analyzing historical flood data and modeling the impact of future flood scenarios, guiding the design and construction of flood-resistant buildings and infrastructure.

Despite the significant potential of AI in flood prediction and mitigation, there are challenges to consider. The accuracy of AI-driven flood prediction systems depends on the quality and availability of data, which can vary across regions. Additionally, the implementation of AI technologies requires investment in infrastructure, data management systems, and skilled personnel. Ensuring data privacy and security is also critical, as flood prediction systems involve sensitive information that could be exploited if not properly protected.

In conclusion, AI offers transformative capabilities for flood prediction and mitigation, providing real-time insights, accurate risk assessments, and effective response strategies. By leveraging AI technologies, authorities can enhance their ability to predict and respond to floods, minimizing the impact on communities and infrastructure. As advancements in AI continue, its applications in flood management are expected to expand, contributing to the resilience and sustainability of our societies in the face of increasing flood risks.

Chapter 7: AI-Enhanced Water Conservation Strategies

Water conservation is essential for ensuring the sustainable use of hydraulic resources and addressing the challenges of water scarcity. AI technologies offer innovative solutions for enhancing water conservation efforts, providing data-driven insights and optimizing water use across various sectors. By leveraging AI, communities, industries, and individuals can implement more effective water-saving strategies, reduce waste, and promote sustainable practices.

One of the primary applications of AI in water conservation is the development of smart water management systems. These systems use a combination of sensors, data analytics, and machine learning algorithms to monitor and manage water usage in real-time. For example, smart meters installed in homes and businesses can continuously track water consumption, providing detailed information on usage patterns. AI algorithms analyze this data to identify inefficiencies and recommend water-saving measures, such as fixing leaks, optimizing irrigation schedules, and adjusting household water use behaviors. By providing personalized recommendations, smart water management systems empower users to take proactive steps towards conserving water.

In addition to residential applications, AI-driven water conservation strategies are being implemented in agriculture, one of the largest consumers of water. Precision agriculture technologies, powered by AI, enable farmers to optimize water use and improve crop yields. Soil moisture sensors, weather data, and satellite imagery are integrated with AI algorithms to determine the precise water needs of crops. This information is used to develop irrigation schedules that minimize water waste and ensure that crops receive adequate moisture. For example, AI-driven systems can detect areas of the field that require more water and adjust irrigation accordingly, preventing over-irrigation and reducing water consumption. By optimizing water use, precision agriculture technologies contribute to sustainable farming practices and enhance food security.

AI is also playing a crucial role in industrial water conservation. Industries, particularly those that require significant amounts of water for production processes, face increasing pressure to reduce their water footprint. AI-driven water management systems enable industries to monitor water use, detect inefficiencies, and implement water-saving measures. For example, machine learning algorithms can analyze data from sensors installed in industrial processes to identify opportunities for water recycling and reuse. By optimizing water use in cooling systems, cleaning processes, and manufacturing operations, industries can significantly reduce their water consumption and minimize environmental impact.

Moreover, AI can enhance water conservation efforts in urban planning and infrastructure development. Urban areas are experiencing rapid population growth, leading to increased demand for water and strain on existing water resources. AI technologies can support the development of sustainable urban water management systems by analyzing data on water consumption, population growth, and climate patterns. For example, AI-driven models can predict future water demand and identify potential water supply challenges. This information enables urban planners to design infrastructure that promotes water conservation, such as rainwater harvesting systems, green roofs, and efficient drainage networks. By integrating AI into urban planning, cities can develop resilient water management strategies that support sustainable growth and ensure reliable water supply.

Furthermore, AI is being used to promote public awareness and education on water conservation. Digital platforms and mobile applications powered by AI provide users with information on water-saving practices and tips for reducing water consumption. These platforms can also gamify water conservation efforts, encouraging users to participate in challenges and track their progress. By raising awareness and promoting behavioral changes, AI-driven tools contribute to a culture of water conservation and empower individuals to make a positive impact.

Despite the significant potential of AI in water conservation, there are challenges to consider. The effectiveness of AI-driven water management systems depends on the availability and accuracy of data, which can vary across regions. Additionally, the implementation of AI technologies requires investment in infrastructure, data management systems, and capacity building. Ensuring data privacy and security is also critical, as water management systems involve sensitive information that must be protected.

In conclusion, AI offers innovative solutions for enhancing water conservation efforts, providing real-time insights, optimizing water use, and promoting sustainable practices across various sectors. By leveraging AI technologies, communities, industries, and individuals can implement more effective water-saving strategies, reduce waste, and promote the sustainable use of hydraulic resources. As advancements in AI continue, its applications in water conservation are expected to expand, contributing to the resilience and sustainability of our societies in the face of increasing water scarcity.

Chapter 8: AI in Wastewater Treatment and Reuse

Wastewater treatment and reuse are crucial for sustainable water management, addressing both environmental and public health concerns. AI technologies are transforming the wastewater treatment industry by enhancing the efficiency, effectiveness, and sustainability of treatment processes. From monitoring and control to predictive maintenance and process optimization, AI-driven solutions are revolutionizing how wastewater is treated and reused.

One of the primary applications of AI in wastewater treatment is real-time monitoring and control. Traditional wastewater treatment plants rely on manual monitoring and control systems, which can be labor-intensive and prone to human error. AI-driven systems use sensors and data analytics to continuously monitor key parameters such as pH levels, chemical concentrations, flow rates, and microbial activity. Machine learning algorithms analyze this data in real-time, identifying patterns and anomalies that indicate potential issues. For example, sudden changes in pH or chemical concentrations can signal a malfunction or contamination event. By providing early warnings and actionable insights, AI systems enable operators to take corrective actions promptly, ensuring the smooth and efficient operation of treatment processes.

AI also plays a crucial role in optimizing the various stages of wastewater treatment. Treatment processes such as aeration, filtration, and chemical dosing require precise control to achieve optimal performance. AI algorithms can model and simulate these processes, identifying the most efficient operational settings and adjusting them dynamically based on real-time data. For instance, AI can optimize aeration systems by adjusting air flow rates to maintain optimal oxygen levels for microbial activity, reducing energy consumption and improving treatment efficiency. Similarly, AI-driven chemical dosing systems can adjust the amount and timing of chemical additions to achieve desired treatment outcomes while minimizing chemical use and costs.

Predictive maintenance is another significant application of AI in wastewater treatment. Treatment plants are complex systems with numerous mechanical and electrical components that require regular maintenance to ensure reliability and prevent failures. AI-powered predictive maintenance systems analyze data from sensors and historical maintenance records to predict when equipment is likely to fail or require servicing. By identifying potential issues before they become critical, these systems enable proactive maintenance scheduling, reducing downtime and repair costs. This approach not only enhances the reliability and longevity of treatment infrastructure but also ensures uninterrupted and efficient operation.

AI technologies also facilitate the reuse of treated wastewater, which is increasingly recognized as a sustainable water management practice. Advanced treatment processes, such as membrane filtration, ultraviolet disinfection, and advanced oxidation, can produce high-quality reclaimed water suitable for various non-potable and even potable uses. AI-driven systems can optimize these advanced treatment processes, ensuring that the treated water meets the stringent quality standards required for reuse applications. Additionally, AI can support the monitoring and management of distribution systems for reclaimed water, ensuring that it is safely and efficiently delivered to end-users.

Despite the significant potential of AI in wastewater treatment and reuse, there are challenges to consider. The implementation of AI technologies requires investment in infrastructure, data management systems, and skilled personnel. The accuracy and reliability of AI-driven systems depend on the quality and availability of data, which can vary across regions and treatment plants. Ensuring data privacy and security is also critical, as wastewater treatment involves sensitive information related to public health and environmental protection. Addressing these challenges requires collaboration between technology providers, policymakers, and water authorities to develop affordable, accessible, and secure AI solutions for wastewater treatment and reuse.

In conclusion, AI offers transformative capabilities for wastewater treatment and reuse, enhancing the efficiency, effectiveness, and sustainability of treatment processes. By leveraging AI technologies, water authorities can improve real-time monitoring and control, optimize treatment operations, implement predictive maintenance, and facilitate the safe reuse of treated wastewater. As advancements in AI continue, its applications in wastewater management are expected to expand, contributing to the sustainable and responsible use of hydraulic resources and promoting environmental and public health.

Chapter 9: AI and Remote Sensing for Water Resource Management

Remote sensing technologies, such as satellites, drones, and aerial imagery, have revolutionized the monitoring and management of water resources. By capturing high-resolution data over large and remote areas, these technologies provide critical insights into hydrological processes, water availability, and environmental changes. AI enhances the capabilities of remote sensing by processing and analyzing vast amounts of data, enabling more accurate and timely water resource management decisions.

One of the primary applications of AI in remote sensing for water resource management is the monitoring of surface water bodies, such as rivers, lakes, and reservoirs. Satellites equipped with multispectral and hyperspectral sensors can capture detailed images of these water bodies, providing information on their extent, volume, and quality. AI algorithms process this imagery to detect changes in water levels, identify areas affected by drought or flooding, and assess the impact of human activities on water resources. For example, AI-driven systems can analyze satellite imagery to monitor the extent of irrigation and its effects on river flows, helping water managers optimize water allocations for agriculture and other uses.

AI also plays a crucial role in the monitoring of groundwater resources, which are vital for drinking water supplies, agriculture, and industrial use. Traditional methods of groundwater monitoring often rely on point measurements from wells, which may not provide a comprehensive picture of groundwater dynamics. Remote sensing technologies, combined with AI algorithms, can estimate groundwater levels and changes over large areas by analyzing surface indicators such as vegetation health, soil moisture, and land subsidence. These AI-driven models provide valuable insights into groundwater recharge rates, depletion trends, and the impacts of land use and climate change on groundwater resources.

In addition to monitoring surface and groundwater resources, AI-enhanced remote sensing supports the assessment of water quality. Satellites and drones equipped with sensors can capture data on various water quality parameters, such as turbidity, chlorophyll concentration, and the presence of harmful algal blooms. AI algorithms analyze this data to detect pollution events, identify sources of contamination, and assess the effectiveness of water quality management interventions. For instance, AI-driven systems can monitor the extent and severity of algal blooms in lakes and coastal areas, providing early warnings to protect public health and aquatic ecosystems.

AI technologies also facilitate the modeling and prediction of hydrological processes, such as precipitation, evapotranspiration, and runoff. Remote sensing data, combined with meteorological and hydrological models, can be used to develop AI-driven models that simulate and predict water cycle dynamics. These models enable water managers to forecast water availability, assess the impacts of climate variability and change, and plan for extreme events such as floods and droughts. For example, AI-driven hydrological models can predict river flows and flood risks based on precipitation forecasts and land use changes, informing the design and implementation of flood mitigation measures.

AI technologies also facilitate the modeling and prediction of hydrological processes, such as precipitation, evapotranspiration, and runoff. Remote sensing data, combined with meteorological and hydrological models, can be used to develop AI-driven models that simulate and predict water cycle dynamics. These models enable water managers to forecast water availability, assess the impacts of climate variability and change, and plan for extreme events such as floods and droughts. For example, AI-driven hydrological models can predict river flows and flood risks based on precipitation forecasts and land use changes, informing the design and implementation of flood mitigation measures.

Despite the significant potential of AI and remote sensing for water resource management, there are challenges to consider. The accuracy and reliability of AI-driven models depend on the quality and resolution of remote sensing data, which can vary based on sensor capabilities and environmental conditions. Data integration and interoperability are also critical, as water resource management requires the synthesis of data from diverse sources and formats. Addressing these challenges requires collaboration between technology providers, water managers, researchers, and policymakers to develop robust, scalable, and accessible AI solutions for remote sensing and water resource management.

In conclusion, AI and remote sensing technologies offer transformative capabilities for the monitoring and management of water resources. By enhancing the analysis of surface and groundwater data, supporting water quality assessment, and enabling the modeling and prediction of hydrological processes, AI-driven remote sensing provides valuable insights for sustainable water resource management. As advancements in AI and remote sensing continue, their applications in water resource management are expected to expand, contributing to the responsible and efficient use of hydraulic resources.

Chapter 10: AI in Stormwater Management

Stormwater management is essential for mitigating the impacts of urbanization, reducing flood risks, and protecting water quality. AI technologies offer innovative solutions for enhancing stormwater management by improving the design, operation, and monitoring of stormwater systems. From predictive modeling to real-time control, AI-driven approaches enable more effective and sustainable management of stormwater in urban and rural areas.

One of the primary applications of AI in stormwater management is the prediction of stormwater runoff and flood risks. Traditional hydrological models often rely on historical data and simplified assumptions, which may not accurately capture the complexities of urban hydrology. AI-driven models, on the other hand, can analyze large datasets from weather forecasts, land use patterns, and drainage infrastructure to predict stormwater runoff and flood risks with greater accuracy. Machine learning algorithms can identify patterns and relationships in the data, providing insights into how different factors contribute to stormwater dynamics. For example, AI models can predict how changes in land use, such as urban development or deforestation, will affect stormwater runoff and flood risks, informing the design of mitigation measures.

AI also enhances the design and optimization of stormwater infrastructure, such as detention basins, green roofs, and permeable pavements. These infrastructure elements are critical for managing stormwater quantity and quality, but their design and operation require careful consideration of local conditions and environmental factors. AI-driven design tools can simulate the performance of different stormwater management practices under various scenarios, identifying the most effective solutions for specific sites. For instance, AI algorithms can optimize the size and placement of detention basins to maximize their flood control benefits while minimizing costs and environmental impacts. Similarly, AI-driven models can assess the effectiveness of green infrastructure, such as bioswales and rain gardens, in reducing stormwater runoff and improving water quality.

Real-time monitoring and control of stormwater systems are another significant application of AI. Traditional stormwater management often relies on static designs and manual operations, which may not respond effectively to changing conditions. AI-driven systems, equipped with sensors and data analytics, can continuously monitor key parameters such as rainfall, water levels, and flow rates in stormwater infrastructure. Machine learning algorithms analyze this data in real-time, identifying trends and anomalies that indicate potential issues. For example, AI systems can detect blockages or malfunctions in drainage networks, enabling timely maintenance and repairs. Additionally, AI-driven control systems can dynamically adjust the operation of stormwater infrastructure, such as opening and closing valves or gates, to optimize performance and reduce flood risks.

AI technologies also support the integration of stormwater management with broader urban planning and environmental protection efforts. By analyzing data from multiple sources, including land use, transportation, and climate models, AI systems provide a comprehensive understanding of how urban development and environmental changes affect stormwater dynamics. This integrated approach enables the identification of synergies and trade-offs between stormwater management and other urban goals, such as reducing heat island effects, improving air quality, and enhancing biodiversity. For example, AI-driven models can assess how the implementation of green infrastructure will affect not only stormwater management but also urban cooling and habitat connectivity, guiding the development of multi-functional and sustainable urban designs.

Community engagement and participation in stormwater management are also facilitated by AI technologies. Digital platforms and mobile applications powered by AI enable citizens to report stormwater issues, such as flooding or pollution, in real-time. These platforms can aggregate and analyze citizen-reported data to provide valuable insights into local stormwater conditions and facilitate timely responses from authorities. AI-driven tools can also educate and empower communities by providing information on stormwater management practices, promoting awareness of flood risks, and encouraging participation in community-based initiatives.

Despite the significant potential of AI in stormwater management, there are challenges to consider. The implementation of AI technologies requires investment in infrastructure, data management systems, and skilled personnel. The accuracy and reliability of AI-driven models depend on the quality and availability of data, which can vary across regions and urban areas. Ensuring data privacy and security is also critical, as stormwater management involves sensitive information related to public safety and environmental protection. Addressing these challenges requires collaboration between technology providers, water authorities, urban planners, and communities to develop robust, secure, and accessible AI solutions for stormwater management.

In conclusion, AI offers transformative capabilities for enhancing stormwater management, improving the prediction of stormwater runoff and flood risks, optimizing the design and operation of stormwater infrastructure, and supporting integrated urban planning. By leveraging AI technologies, water authorities and urban planners can develop more effective and sustainable stormwater management strategies, reducing flood risks and protecting water quality. As advancements in AI continue, its applications in stormwater management are expected to expand, contributing to the resilience and sustainability of urban environments.